Standard of Ministry of Water Resources of
the People's Republic of China

SL 279—2016

Replace SL 279—2002

Specification for Design of Hydraulic Tunnels

Drafted by:
China Water Northeastern Investigation, Design and Research Co., Ltd.

Translated by:
China Water Northeastern Investigation, Design and Research Co., Ltd.

China Water & Power Press
Beijing 2019

图书在版编目（ＣＩＰ）数据

水工隧洞设计规范：SL 279-2016 = Specification for Design of Hydraulic Tunnels : SL 279-2016 : 英文 / 中华人民共和国水利部发布. -- 北京 : 中国水利水电出版社, 2019.4
 ISBN 978-7-5170-7627-8

Ⅰ. ①水… Ⅱ. ①中… Ⅲ. ①水工隧洞－设计规范－中国－英文 Ⅳ. ①TV672-65

中国版本图书馆CIP数据核字(2019)第074485号

书　　名	Specification for Design of Hydraulic Tunnels SL 279—2016
作　　者	中华人民共和国水利部　发布
出版发行	中国水利水电出版社 （北京市海淀区玉渊潭南路1号D座　100038） 网址：www.waterpub.com.cn E-mail: sales@waterpub.com.cn 电话：（010）68367658（营销中心）
经　　售	北京科水图书销售中心（零售） 电话：（010）88383994、63202643、68545874 全国各地新华书店和相关出版物销售网点
排　　版	中国水利水电出版社微机排版中心
印　　刷	清淞永业（天津）印刷有限公司
规　　格	140mm×203mm　32开本　2.625印张　71千字
版　　次	2019年4月第1版　2019年4月第1次印刷
定　　价	160.00元

凡购买我社图书，如有缺页、倒页、脱页的，本社营销中心负责调换

版权所有·侵权必究

Introduction to English Version

Department of International Cooperation, Science and Technology of Ministry of Water Resources, P. R. China (hereinafter DICST) has the mandate of managing the formulation and revision of water technology standards in China.

Translation of this standard from Chinese into English is organized by DICST in accordance with due procedures and regulations applicable in China.

The English version of this standard is identical to its Chinese original SL 279—2016 *Specification for Design of Hydraulic Tunnels*, which was formulated and revised under the auspices of DICST.

Translation of this standard is undertaken by China Water Northeastern Investigation, Design and Research Co., Ltd.

Translation task force includes MA Jun, SONG Shouping, LIU Jie, ZHANG Xiaotao, ZHANG Yu, YIN Yiguang, CHEN Quanbao, LIU Wenbin, MA Donghe, CHANG Yuan, YANG Yan, ZHANG Jianhui, QI Liwei, LI Huan and ZHENG Yuling.

This standard is reviewed by JIN Hai, GUO Jun, MENG Zhimin, WU Nongdi, JIN Feng, SUN Feng, CHEN Shangfa, BAO Shujun, CHEN Shaosong, YANG Haiyan, CHEN Liqiu, YANG Weijiu and YANG Kete.

Department of International Cooperation, Science and Technology
Ministry of Water Resources, P. R. China

Foreword

According to the schedule of the formulation and revision of water technology standards and the requirements of SL 1—2014 *Specification for the Drafting of Technical Standards of Water Resources*, SL 279—2002 *Specification for Design of Hydraulic Tunnel* has been revised.

This standard is composed of 11 chapters and 4 annexes. The main contents include the following:

——Basic data.

——Layout of tunnels.

——Pressure status, shape and size of tunnels.

——Hydraulic design of tunnels.

——Design of soil tunnels.

——Design of tunnels in poor ground.

——Tunnel supports and linings.

——Grouting, seepage control and drainage of tunnels.

——Tunnel operation and maintenance.

The major revisions are as follows:

——Adding some terms and symbols.

——Adding relevant contents of excavation by tunnel boring machines.

——Adding the calculation of bearing capacity limit states and checking of serviceability limit states for reinforced concrete lining structures.

——Cancelling relevant design contents of crack resistance for reinforced concrete lining structures.

——Modifying the calculation method of sliding stability for

a plug.

——Adding the calculation of head loss for hydraulic tunnels.

——Adding the calculation of crack width for concrete linings.

——Cancelling the structural calculation of lining for a circular pressure tunnel.

——Cancelling the structural calculation of grouted prestressed concrete linings.

The compulsory provisions Clauses 1 and 2 of Article 5.1.2, Articles 9.8.8 and 10.1.1 are printed in bold type and must be enforced strictly.

The previous standards which are substituted by this standard are the following:

—— SD 134—84

—— SL 279—2002

This standard is approved by Ministry of Water Resources of the People's Republic of China.

This standard is explained by General Institute of Water Resources and Hydropower Planning and Design of Ministry of Water Resources of the People's Republic of China.

This standard is drafted by China Water Northeastern Investigation, Design and Research Co., Ltd.

This standard is published and distributed by China Water & Power Press.

Chief drafters of this standard are JIN Zhenghao, SONG Shouping, FAN Jingchun, ZHANG Jianhui, ZHENG Jun, YU Shengbo, LI Huan, CHENG Yujiao, LYU Hongfei, LIU Yang, WANG Chao, FU Xin, LIU Shan, ZHENG Yuling, LIU Feng, DONG Yanchao, JIANG Shuli, LI Jun, WANG

Chen and GU Yixin.

The leading expert of the technical review meeting of this standard is WEN Xuyu.

The format examiner of this standard is ZHANG Ping.

Contents

Introduction to English Version
Foreword
1 General Provisions 1
2 Terms and Symbols 3
   2.1 Terms 3
   2.2 Symbols 4
3 Basic Data 6
4 Layout of Tunnels 8
   4.1 Selection of Tunnel Alignment 8
   4.2 Layout of Tunnel Inlets and Outlets 13
   4.3 Layout of Multipurpose Tunnels 15
5 Pressure Status, Shape and Size of Tunnels 17
   5.1 Selection of Pressure Status 17
   5.2 Cross-Sectional Shape 18
   5.3 Cross-Sectional Size 19
6 Hydraulic Design of Tunnels 22
   6.1 Principle of Calculation 22
   6.2 Cavitation Damage Prevention Design of
        High-Velocity Tunnels 23
7 Design of Soil Tunnels 25
   7.1 Supports and Linings of Soil Tunnels 25
   7.2 Joints, Seepage Control and Waterstops for the
        Linings of Soil Tunnels 27
8 Design of Tunnels in Poor Ground 29
9 Tunnel Supports and Linings 34
   9.1 General 34

9.2	Loads and Load Combinations	35
9.3	Unreinforced- and Reinforced-Concrete Linings	38
9.4	Prestressed Concrete Linings	39
9.5	Unlined and Bolt-Shotcrete Lined Tunnels	41
9.6	Design of Reinforced Concrete Lined Bifurcation	45
9.7	Joints in Linings	46
9.8	Design of Water-Retaining Plugs	47

10 Grouting, Seepage Control and Drainage of Tunnels 50
   10.1 Grouting 50
   10.2 Seepage Control and Drainage 51

11 Tunnel Operation and Maintenance 53

Annex A   Head Loss Calculation for Hydraulic Tunnels 54
   A.1 Friction Losses 54
   A.2 Form Losses 55

Annex B   Cavitation Damage Prevention Design of High-Velocity Tunnels 65

Annex C   Calculation Method and Reduction Coefficient of External Water Pressure 68

Annex D   Calculation of Concrete Lining Crack Width 70

Explanation of Wording 74

1 General Provisions

1.0.1 This specification is formulated to standardize the design of hydraulic tunnels and ensure the design quality with the view of making use of the state-of-the-art technology, meeting the requirements for safety, applicability, cost-effectiveness, and rationality.

1.0.2 This specification is applicable to the design of Class 1, Class 2 and Class 3 hydraulic tunnels in water resources and hydropower projects, but not applicable to the design of hydraulic tunnels with steel linings in rock/soil mass.

1.0.3 The classification of hydraulic tunnels shall be in accordance with GB 50201 *Standard for Flood Control* and SL 252 *Standard for Rank Classification and Flood Protection Criteria of Water and hydropower Projects*. After justification, the class of tunnels may be:

    1 Increased by one class for Class 2 and Class 3 tunnels with extraordinarily complex geological conditions, or extraordinarily high water head and flow velocity, or whose failure may cause huge losses.

    2 Lowered by one class for low-head or low-velocity tunnels whose failure will not cause huge losses.

1.0.4 The safety monitoring design of hydraulic tunnels shall meet the requirements of SL 725 *Design Specification for Safety Monitoring in Water and Hydropower Projects*.

1.0.5 The design of hydraulic tunnels shall satisfy the requirements of the master plan and environmental protection.

1.0.6 The standards quoted in this standard mainly include:

    GB/T 21120 *Synthetic Fibres for Cement, Cement Mortar*

and Concrete

    GB 50086   *Technical Code for Engineering of Ground Anchorages and Shotcrete Support*

    GB/T 50145   *Standard for Engineering Classification of Soil*

    GB 50201   *Standard for Flood Control*

    GB 50487   *Code for Engineering Geological Investigation of Water Resources and Hydropower*

    SL 62   *Technical Specification for Cement Grouting of Hydraulic Structures*

    SL 74   *Design Code for Steel Gate in Water Resources and Hydropower Projects*

    SL 191   *Design Code for Hydraulic Concrete Structures*

    SL 203   *Specifications for Seismic Design of Hydraulic Structures*

    SL 252   *Standard for Rank Classification and Flood Protection Criteria of Water and Hydropower Projects*

    SL 253   *Design Code for Spillway*

    SL 285   *Design Specification for Intake of Hydraulic and Hydroelectric Engineering*

    SL 377   *Technical Specification of Shotcrete and Rock Bolt for Water Resources and Hydropower Project*

    SL 725   *Design Specification for Safety Monitoring in Water and Hydropower Projects*

    DL/T 5207   *Technical Specification for Abrasion and Cavitation Resistance of Concrete in Hydraulic Structures*

1.0.7 The design of hydraulic tunnels shall comply with provisions stipulated not only in this standard but also in other prevailing national/ministerial/sectoral standards.

2 Terms and Symbols

2.1 Terms

2.1.1 Hydraulic tunnel

An elongated opening excavated inside the mountain or under the ground for the purpose of water conveyance.

2.1.2 Pressure tunnel

A hydraulic tunnel full of water and subject to its pressures on the tunnel wall.

2.1.3 Free-flow tunnel

A hydraulic tunnel partially filled with water flow which has free surface.

2.1.4 Tunnel lining

A lining structure required for stabilizing the tunnel's surrounding rock and providing good hydraulic conditions.

2.1.5 Tunnel support

Installation of structures, elements, etc., to reinforce the tunnel's surrounding rock.

2.1.6 Unlined tunnel

A hydraulic tunnel of which most inner surface is not lined.

2.1.7 Pattern bolts

Bolts installed on the overall surface at a certain spacing and in a regular pattern as required for rock mass stability.

2.1.8 Backfill grouting

Injection of fluid grout to fill empty spaces and voids between tunnel lining and excavated rock or soil surface, subsurface voids, in order to strengthen the integrity of the structure or foundation.

2.1.9 Consolidation grouting

Injection of fluid grout to consolidate the surrounding rock with geological defects such as fissures or fractured zones, in order to strengthen integrity and bearing capacity of the rock.

2.1.10 Chemical grouting

Injection of fluid grout predominantly composed of chemical materials.

2.1.11 Tunnel transition section

A connection where the cross-sectional shape or size of the tunnel gradually alters.

2.1.12 Convergent deformation

The measured displacement between two observation points fixed on the surface of a tunnel.

2.2 Symbols

2.2.1 Geometric parameters

- A_i ——Contact area of bottom and sides of a plug with rock mass/concrete, except that of crown;
- C_{RM} ——Minimum rock cover;
- α ——Slope angle of valley;
- b ——Excavated tunnel width;
- h ——Excavated tunnel height.

2.2.2 Physical and mechanical parameters

- C' ——Cohesion of shear breaking strength between concrete and rock mass or between concrete and concrete;
- f' ——Friction coefficient of shear breaking strength between concrete and rock mass or between concrete and concrete;
- q_h ——Uniformly distributed horizontal rock pressure;
- q_v ——Uniformly distributed vertical rock pressure;
- r_R ——Unit weight of rock;
- r_w ——Unit weight of water.

2.2.3 Actions and action effects

h_s ——Hydrostatic pressure head in a tunnel;

K ——Safety factor;

F ——Empirical coefficient;

R ——Design value of bearing capacity;

S ——Design value of load effect;

$\sum P$ ——Sum of tangential components of all loads borne by the plug on the sliding plane;

$\sum W$ ——Sum of normal components of all loads borne by the plug on the sliding plane;

λ_i ——Effective area coefficient of contact surface of bottom and sides of a plug with rock mass/concrete except that of crown.

3 Basic Data

3.0.1 According to the requirements of project layout, purposes of tunnel, construction conditions and design stage, the following basic data shall be collected for the design of a hydraulic tunnel:

—River basin master plan, project purposes, project layout, characteristic stage of reservoir/ river course, discharge requirements for the tunnel, operation modes, rules of water withdrawal from river course, etc.

—Geological data of the project area and basic seismic intensity.

—Relevant hydrological data, meteorological data and results of hydrological design, information on construction materials and construction methods, mechanical and electrical equipment and surge/relief facilities, penstocks, gates, valves, etc.

—The requirements of environmental protection, soil and water conservation, historical and cultural relics, mineral resources in tunnel area, etc.

3.0.2 The following basic geological data in tunnel area shall be obtained depending on different design phases of hydraulic tunnels:

—Engineering geological conditions along the tunnel alignment, such as boundaries between different rock/soil types, attitude, characteristics, major discontinuities, classification and major physical and mechanical parameters of surrounding rock mass.

—Hydrogeological conditions along the tunnel alignment, such as groundwater table, water temperature and chemical composition, in particular, the source of water gushing like the aquifer, karst cave, pervious zones, and faults and fractured zones which are connected with the creeks and gullies, etc.

—Tunneling conditions of inlets and outlets, and stability of cut slopes at the portal.

—In situ stresses, ground temperature, rock bursts, hazardous gases and radioactive elements, etc.

—Estimate of adverse geological conditions.

3.0.3 The geological exploration for a hydraulic tunnel shall comply with GB 50487.

3.0.4 For Class 1 and Class 2 hydraulic tunnels and those with adverse geological conditions along the tunnel alignments, based on different requirements of each design phase, the representative locations shall be selected to conduct related tests and observations. Designers, together with geologists, shall propose the requirements of tests and observations according to the design requirements and relevant criteria.

3.0.5 The classification of the surrounding rock mass of a hydraulic tunnel shall comply with GB 50487 for rock tunnels and GB/T 50145 for soil tunnels.

3.0.6 During the excavation of long tunnels at great depth, the geological forecasts/predictions or advance exploration shall be intensified, and the design parameters shall be timely adjusted or modified accordingly.

3.0.7 After commencement of hydraulic tunnel excavation, the designers shall have a full knowledge of the variations of geological conditions in every portion of tunnels so as to recheck, supplement or modify the design in time. In the event of the geologic problems which might compromise the safety of construction and operation, specific studies shall be made.

4　Layout of Tunnels

4.1　Selection of Tunnel Alignment

4.1.1　Hydraulic tunnel alignment shall be selected through technical and economic comparison in line with the purposes and characteristics of tunnels, with comprehensive consideration of various factors including topography, geology, ecological environment, soil and water conservation, layouts of project and structures along the tunnel alignment, hydraulics, construction and transport, operation, etc.

4.1.2　In addition to meeting the requirements of general layout of the project, the alignment should be located in an area with simple geological structures, stable and intact rock mass, favorable hydrogeological conditions and construction convenience. And the following requirements shall be met:

1　There should be a large angle between the tunnel alignment and the strike of strata, tectonic discontinuities and major weakness zones. The angle should not be less than 30° in the case of intact blocky rock mass or tightly-cemented, thick-layered rock mass consisting of hard and intact rocks. It should not be less than 45° in the case of thin-layered rock mass, especially the one consisting of steep and thick layers with poor bonding between them.

2　When passing through major tectonic zones, the hydraulic tunnel alignment shall be determined through technical and economic comparison on the basis of the influence of the adverse geological structures and their combinations on the surrounding rock stability, with due consideration of various factors including construction, operation, construction duration, cost, etc.

3　For the tunnel with faults and cleavages, adverse tectonic planes, weakness zones, alteration zones, swelling rocks encountered, the influence of groundwater movement on the stability of the rock mass shall be considered. The gully which is a potential source of strong water recharge should be avoided in selecting the alignment of tunnel.

  4 Tunnel alignment should avoid well developed karst areas.

  5 The angle between the tunnel alignment and the direction of maximum horizontal in situ stress should be as small as possible in the areas with high in situ stress.

4.1.3 When selecting a tunnel alignment, potential unstable rock mass in local areas shall be analyzed and predicted, so that appropriate measures can be proposed accordingly.

4.1.4 The minimum rock cover of a tunnel in vertical and lateral directions, see Figure 4.1.4, shall be analyzed and determined based on such factors as topography, geological conditions, confinement and permeability of the rock mass, internal water pressure, lining types, and the following shall be met:

  1 For inlet, outlet and free-flow tunnel, when the reasonable construction procedures and engineering measures have been taken to ensure the safety during construction and operation, the minimum rock cover may be free from any requirement.

  2 For a pressure tunnel, the minimum rock cover may be calculated by Equation (4.1.4), and the analysis shall be carried out by using finite element method if necessary:

$$C_{RM} = \frac{F\gamma_w h_s}{\gamma_R \cos\alpha} \quad (4.1.4)$$

Where:

  C_{RM} = Minimum rock cover (excluding the overburden of the completely and intensively weathered rock), m;

  h_s = Hydrostatic pressure head in tunnel, m;

γ_w = Unit weight of water, kN/m³;
γ_R = Unit weight of rock, kN/m³;
α = Slope angle, (°), if $\alpha > 60°$, $\alpha = 60°$ is taken;
F = Empirical coefficient, 1.3 – 1.5 may be taken.

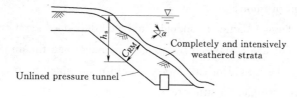

Figure 4.1.4 **Rock cover of a pressure tunnel**

3 For a pressure tunnel, the minimum rock cover shall prevent failure caused by excessive leakage and hydraulic jacking of the rock mass from occurring, and the seepage gradient of the surrounding rock shall satisfy the stability requirements.

4 High – pressure tunnel design shall not only comply with Clauses 2 and 3 of this article but also satisfy the requirement that the maximum internal water pressure is less than the minimum in situ stress, and analysis shall be carried out using finite element method if necessary.

4.1.5 The net spacing between adjacent tunnels shall be determined through analysis based on arrangement needs, geological conditions, stress and deformation of the surrounding rock mass, cross-sectional shape and size of tunnels, construction methods and operating conditions, etc. It shall be such as to guarantee that failure of rock mass between tunnels caused by seepage and hydraulic jacking will not occur in operation, and the rock thickness should not be less than two times the excavated tunnel diameter or width. If it is necessitated by project layout, the net spacing may be reduced appropriately after justification, but it shall in no case be less than the diameter or width of excavated tunnel.

4.1.6 The tunnel alignment should be such as to avoid the adverse influence on adjacent structures. For a tunnel passing through dam foundation, dam abutment or other structure foundation, the spacing between the base of the structure and the tunnel shall meet the requirements of structure, seepage control, etc.

4.1.7 When a tunnel encounters gullies, the alternative of bypassing or crossing the gullies may be selected through technical and economic comparison based on topography, geology, hydrology and construction conditions. When an alternative of crossing the gullies is adopted, the type and location of the crossing shall be reasonably selected, and intensified engineering measures shall be required for unstable gully slopes and the connecting parts of crossing structure and tunnel, etc.

4.1.8 A soil tunnel near a river or mountainside shall avoid eccentric compression and water erosion of the mountain slope or landslides.

4.1.9 A tunnel alignment should be a straight line in plan. The following requirements shall be met when a curved alignment is inevitable:

1 When a curved alignment is adopted for a low-velocity free-flow tunnel, the curve radius should not be less than five times the tunnel diameter or width, and the deflection angle should not be more than 60°. For a low-velocity pressure tunnel, these restrictions may be relaxed. In this case, the curve radius should not be less than three times the tunnel diameter or width, and the deflection angle should not be more than 60°.

2 A curved stretch shall be avoided in the design of a high-velocity free-flow tunnel. When a curved stretch is set for a high-velocity pressure tunnel, the curve radius and the deflection angle shall be determined through hydraulic model tests.

3 A straight stretch shall be provided at each end of the curved stretch, and should not be less than five times the tunnel diameter or width in length.

4.1.10 When a vertical curve alignment is adopted for the main part of tunnel, the type of high-velocity tunnel and the radius of the vertical curve shall be determined through hydraulic model tests. The radius of the vertical curve of a low-velocity free-flow tunnel should not be less than five times the tunnel diameter or width, and the requirements may be appropriately relaxed for a low-velocity pressure tunnel.

4.1.11 When a horizontal curve or a vertical curve is adopted in hydraulic tunnels, the requirements of construction methods, limits of construction capability and large construction equipment shall be considered.

4.1.12 The longitudinal slope of a tunnel may be determined by operating requirements, hydraulic conditions, foundation elevation of structures along the alignment, upstream and downstream connections, construction and maintenance conditions, etc. , and the following shall be met:

1 The requirement of non-silting velocity shall be met.

2 The longitudinal slope along the alignment should not be changed frequently.

3 The arrangement of flat or reverse slopes should be avoided. But when such arrangement is required, provision of maintenance and drainage conditions shall be considered.

4 In the case of a long tunnel for irrigation and water supply or conveyance, the arrangement of water diversion or water withdrawal facilities along the tunnel shall be considered when determining its longitudinal slope.

4.1.13 For a silt-flushing tunnel, its horizontal and vertical curve, deflection angle and longitudinal slope grade should be determined through hydraulic model tests.

4.1.14 When adits are designed for a tunnel, the number and lengths of the adits shall be determined through technical and economic comparison based on the topographical and geological conditions, access, tunneling work of the tunnel between adits, easy mucking and construction duration requirement. When the geological condition is adverse, the influence of adits on the tunnel shall be studied.

4.1.15 When a tunnel boring machine (TBM) is used, the tunnel alignment should avoid those areas geologically unsuitable for it.

4.2　Layout of Tunnel Inlets and Outlets

4.2.1 The layout of tunnel inlets and outlets shall be determined based on such factors as general layout of the project, topographical and geological conditions. In addition, the following shall be met:

1 The function and operating safety requirements shall be met.

2 The layout shall make the streamline smooth, inflow uniform, and outflow stable.

3 The requirements of siltation control, ice control, erosion control, and trash prevention shall be met.

4 The arrangement of gates, trash rack, cleaning equipment and the requirements of access shall be considered.

4.2.2 Inlets and outlets should be located in the areas with simple geological structures, sound rock mass, shallow weathered or unloading zones. The arrangement should avoid those areas with poor geological structures, gullies, and those prone to collapse, landslides, debris flows, etc.

4.2.3 High slope excavation should be avoided at inlet and outlet portals as much as possible. Appropriate reinforcement, water control and drainage measures shall be taken based on the stability analysis of the excavated slope.

4.2.4 The necessary range of slope cleaning shall be defined for inlet and outlet portals, and the appropriate engineering measures shall be taken to protect the normal operation against the sliding of overburden and falling of loose rock blocks under the actions of wind, water flow, waves, variations of water level, earthquake, etc.

4.2.5 The portal of a soil tunnel shall be located in areas with stable slopes and sound soil conditions. The design slope of a soil tunnel portal shall be analyzed and determined by slope stability analysis based on soil properties and excavation height.

4.2.6 Permanent joints shall be provided at the connection position between the soil tunnel portal and other structures such as aqueducts and rock tunnels. In cold or frigid region, the tunnel portal foundation shall be deep enough to resist freezing.

4.2.7 The outlet of a water releasing tunnel shall be designed as per the following requirements:

1 The outlet sectional area of a pressure water releasing tunnel should be contracted to 85% to 90% of the cross-sectional area of the tunnel. If the geometry varies greatly along the tunnel and the flow condition in the tunnel is adverse, the area should be contracted to 80% to 85%. The descending of tunnel roof should be used for contraction, and for important tunnels, the design should be verified through hydraulic model tests.

2 The outlet transition section configuration of a pressure water releasing tunnel should be determined on the basis of flow conditions, service gates type and layout, and the gate opening mode.

3 To make outlet flow well connected with downstream flow, the outlet bottom slope of a pressure water releasing tunnel should be gentle, and lateral diffusion should be smooth. Hydraulic model tests shall be performed when drastic enlargement or drop is adopted. Appropriate measures of outflow guidance should be taken to prevent the outflow from rushing into the main stream when the outlet is close to the main channel/main stream.

4 Alternative measures of energy dissipation and scouring control shall be selected through technical and economic comparison, based on topographical and geological conditions, hydraulic conditions, operating modes, the water depth downstream and variation, scouring resistance capacity of downstream channel, flow connection, the requirements of energy dissipation and scouring control, and the influence on the adjacent structures. In the meanwhile, concerned requirements of SL 253 shall be met.

4.2.8 For air replenishment during dewatering or air exhaust during water filling in a pressure tunnel, ventilation above the water surface in a free-flow tunnel and other tunnel portions needing ventilation, necessary aeration area shall be calculated. The aeration area of a pressure tunnel may be calculated according to SL 74.

4.2.9 The design for intake of hydropower station shall comply with SL 285.

4.3 Layout of Multipurpose Tunnels

4.3.1 In selection of tunnel layout alternatives, the feasibility, rationality and economics of combining temporary and permanent functions and using one tunnel for multiple purposes shall be studied based on the purposes, operating and construction conditions of a tunnel.

4.3.2 For a tunnel with combination of temporary and permanent functions, comprehensive comparison and study of tunnel alignment, longitudinal slope, cross-section, support and lining type, inlet and outlet elevation and locations, operating and maintenance conditions shall be carried out.

4.3.3 For a main tunnel serving both flood releasing and power generation purposes, the following requirements shall be met:

1 Individual operating requirements and good hydraulic conditions shall be satisfied.

2 Flood releasing tunnel should be designed as the main tunnel, and the power tunnel as the branch.

3 Bifurcation type should be determined based on water head, discharge and split ratio. Where necessary, it shall be verified through hydraulic model tests.

4 The length of power tunnel after bifurcation should not be less than ten times the tunnel diameter or width, and may be properly reduced when flood releasing and power generating does not occur simultaneously or pressure control facilities are available in the power intake system.

5 For the main tunnel used for discharging flood, the cross-sectional area of the outlet should not be larger than 85% of that of the main tunnel. For the branch tunnel used for discharging flood, the cross-sectional area of the outlet should not be larger than 70% of that of the branch tunnel.

4.3.4 The possibility of using all or part of a tunnel as a permanent tunnel should be studied in the design of a river diversion tunnel.

4.3.5 When energy dissipation inside the tunnel is adopted for flood releasing tunnel, hydraulic model tests shall be carried out.

4.3.6 The feasibility of using geological exploration adits for other purposes should be considered in the design.

5 Pressure Status, Shape and Size of Tunnels

5.1 Selection of Pressure Status

5.1.1 The pressure status of hydraulic tunnels should be determined through technical and economic comparison based on the functional requirements and characteristics of tunnels, with comprehensive consideration of various factors including topography, geology, general layout of project, hydraulics, construction, operation, etc.

5.1.2 Flow regime in a tunnel shall meet the following requirements:

1 **For a pressure tunnel, alternation between open channel flow and pressure flow shall not be permitted, and the minimum pressure head along the whole tunnel roof shall not be less than 2.0 m under the most unfavorable operating conditions.**

2 **For a high-velocity flood releasing tunnel, alternation between open channel flow and pressure flow shall not be permitted.**

3 For a low-velocity flood releasing tunnel designed with open channel flow under the normal operating condition, alternation between open channel flow and pressure flow may be allowed in case of check flood level.

4 For the outlet section of an open channel flow tunnel, alternation between open channel flow and pressure flow may be allowed for a short period in flood season.

5.1.3 For a river diversion tunnel, the operating mode of alternation between open channel flow and pressure flow may be adopted if it is confirmed after justification the flow regime will not cause the failure of the tunnel under the design flow condition.

5.1.4 Free-flow should be ensured in a soil tunnel in most cases. In the event of pressure flow, the lining type shall be properly selected in terms of such factors as soil resistance, internal and external water pressure and seepage deformation of soil.

5.2 Cross-Sectional Shape

5.2.1 The cross-sectional shape shall be selected as per the following requirements:

1 For pressure tunnel, a circular section should be adopted. If the stability condition of surrounding rock is sound, and the internal and external water pressure are not high, other cross-sectional shapes convenient for construction may be adopted.

2 For free-flow tunnel, an inverted U-shaped section should be adopted. If the geological conditions are adverse, circular or horseshoe-shaped section may be selected.

3 For the inverted U-shaped section, the central angle of the crown should be ranged from 90° to 180°. If the thrust of abutment needs to be increased, the central angle may be less than 90°. The height-width ratio of the section shall be selected according to the hydraulic and geological conditions, and should be in the range of 1.0 to 1.5. A higher ratio should be adopted if water level in the tunnel changes greatly.

5.2.2 If a noncircular cross section is adopted in a high in situ stress area, the height-width ratio of the section shall be governed by in situ stress condition. A low and wide section should be applied if the horizontal in situ stress is greater than the vertical one, and a high and narrow section should be applied if the vertical in situ stress is greater than the horizontal one.

5.2.3 For multipurpose tunnels serving due functions of power generation and flood discharge, river diversion and power generation, or river diversion and flood discharge, etc., the cross-sectional shape shall be determined through technical and economic comparison, and should be verified through hydraulic model tests if necessary.

5.2.4 A long tunnel may have different cross-sectional shapes and lining types, but its variations should not be frequent, and shall meet the following requirements:

1 A transition shall be provided between two different types of cross-sections or lining and its boundary shall be designed as a gentle curve and be convenient for construction.

2 The conical angle of pressure tunnel transition should be in the range of 6° to 10°, and the smaller value should be adopted for the transition with two-way flow. The length of transition should not be less than 1.5 times the tunnel diameter or width.

3 The transition shape of a tunnel with high-velocity free-flow shall be determined through hydraulic model tests.

5.3 Cross-Sectional Size

5.3.1 The cross-sectional size of a tunnel shall be determined by:

1 Trade-off studies for a water conveyance tunnel of a hydropower station or pumping station.

2 The elevations of inlet and outlet of the tunnel and the design or increased discharge for a tunnel of a water transfer project.

3 The requirements of the flow capacity under various operating conditions for a flood releasing tunnel.

4 The requirements of diversion discharge, elevation of inlet, height of cofferdams, connection with outlet flow, etc., for a river diversion tunnel.

5.3.2 For a multipurpose tunnel, the cross-sectional size shall meet various operating requirements, and the size of the shared portion shall be determined through technical and economic comparison.

5.3.3 The minimum cross-sectional size of a tunnel shall meet the following requirements:

1 For tunnels excavated by drilling and blasting method, the inner diameter of circular section should not be less than 2.0 m, and the non-circular section should not be less than 1.8 m in height nor 1.5 m in width.

2 For tunnels excavated by TBM, the minimum size shall meet the requirements of the equipment.

5.3.4 The cross-sectional size of a low-velocity tunnel with open channel flow shall meet the following requirements:

1 For steady flow, if the ventilation conditions are good, the cross-sectional area above the water surface in the tunnel should not be less than 15% of the tunnel area, and the height above the water surface shall not be less than 0.4 m.

2 For unsteady flow, the values in Clause 1 of Article 5.3.4 may be reduced properly if the surge wave has already been taken into account.

3 For a tunnel longer than 1000 m, unlined, or bolt-shotcrete supported, the values in Clause 1 of Article 5.3.4 may be increased properly.

4 For tunnels with requirements such as navigation, the bending radius and angle, the flow area and the area above water surface shall comply with relevant standards.

5.3.5 The cross-sectional size of a high-velocity tunnel with open channel flow shall meet the following requirements:

1 It should be determined through hydraulic model tests in consideration of aeration effects.

2 The area above the aerated water surface should be 15% to 25% of the total tunnel area.

3 For the inverted U-shaped section, the water surface should not be above the vertical wall.

4 The wave peak shall not be above the vertical wall when there are shock waves in the water flow.

5.3.6 The size of service tunnels for TBM shall meet the requirements of assembly and disassembly, and safe passage during construction period. Engineering measures shall be taken to meet the requirements of tunnel operation.

6 Hydraulic Design of Tunnels

6.1 Principle of Calculation

6.1.1 The following shall be calculated selectively based on purposes and different design stages of tunnels:

—Flow capacity;

—Connection between tunnel flow and upstream/downstream flow;

—Head loss;

—Hydraulic grade line;

—Water surface profile;

—Factors regarding aeration, water filling and emptying, and other hydraulic phenomena.

6.1.2 The calculation of friction and form losses of hydraulic tunnels shall meet the following stipulations:

1 The roughness of tunnel lining used for friction loss calculation may be selected from Annex A and based on lining type, tunnel excavation method, changes after operation, etc.

2 The coefficients used for form loss calculation may be selected from Annex A and determined through tests if necessary.

6.1.3 The flow capacity shall be calculated as per:

1 Pipe flow for pressure tunnels.

2 Uniform flow for long free-flow tunnels and non-uniform flow for short ones.

3 Weir flow for free-flow tunnels with surface inlets and pipe flow for those with deep submerged inlets.

6.1.4 In the calculation of water surface profile of free flow tunnels, the type of water surface profile shall be identified firstly, and the calculation may be carried out by methods such as piecewise summation after the control section is selected.

6.1.5 For hydraulic tunnels with high-velocity, large discharge and complicated flow conditions, integral or localized hydraulic model tests shall be carried out so as to justify the hydraulic calculation and structure layout.

6.2 Cavitation Damage Prevention Design of High-Velocity Tunnels

6.2.1 For a high-velocity hydraulic tunnel, the incipient cavitation index at the lowest pressure point or potential point of the selected shape shall be less than the flow cavitation index at this point; otherwise corresponding measures shall be taken. The likelihood of the cavitation damage shall be judged according to Annex B.

6.2.2 The following positions subject to cavitation damage shall be highlighted in the design of a high-velocity hydraulic tunnel:

—The inlet, gate slot, transition section, bifurcation section, bending section, outlet of a pressure tunnel and positions where the flow boundary drastically changes.

—The curve section with steep slope, invert curve section, expansion or contraction section, piers, gate slot and outlet section, etc., of a free-flow tunnel.

—Position of outlet energy dissipation.

6.2.3 For the positions subject to cavitation damage, the following measures should be taken to prevent erosion.

1 Suitable shape should be selected.

2 Localized surface irregularities of flow boundary shall be controlled, and its criterion shall be determined on the basis of Annex B.

3 Aeration should be required. The type, size and location of aeration facilities may be determined through sectional model tests, or based on prototype observation data from completed projects.

    4 Cavitation-resistant materials should be adopted.

    5 Reasonable operating modes should be selected.

6.2.4 For a sediment-laden river, the materials with good corrosion and abrasion resistance shall be adopted by considering the combined effect of abrasion and cavitation on the boundary caused by sand-carrying flow.

7 Design of Soil Tunnels

7.1 Supports and Linings of Soil Tunnels

7.1.1 The design of soil tunnels shall meet the following requirements:

1 For longer tunnels, technical and economic comparison between conventional and shield construction method should be carried out.

2 For soil tunnels, the cross-section should be circular-shaped or horseshoe-shaped.

3 For soil tunnels, composite linings consisting of shotcrete or bolt-shotcrete and reinforced concrete should be adopted. High performance wet shotcrete should be used. The lining should be of integral structure. And structural measures shall be taken to prevent exfiltration.

4 Surface water and construction water in tunnels shall be properly diverted and drained.

5 For a soil tunnel connected with a rock tunnel, its support and lining shall extend into the rock tunnel for a sufficient length. The minimum overburden of rock tunnel at the transition shall not be less than the tunnel diameter.

7.1.2 The surrounding rock pressure (load) on the lining structure of the soil tunnel shall be determined as follows:

1 For a soil tunnel which can form a ground arch, surrounding rock pressure may be estimated according to equilibrium theory of loose medium.

2 For a shallow soil tunnel which cannot form a ground arch, surrounding rock pressure should be calculated according to the gravity of overlying soil above the arch and adjusted based on topographical conditions and stabilization measures adopted in construction.

  3 For a deep soil tunnel which cannot form a ground arch, surrounding rock pressure should be studied specially.

  4 Swelling pressure shall be considered for swelling soil, and its value may be studied and determined through sampling tests or field tests.

  5 For a tunnel portion with underground water, the loads acting on lining structure shall be determined by the joint action of soil and water pressure. For a tunnel portion with high external water pressure, if the soil mass cannot maintain its stability after being supported, measures including drainage and strengthening support or reinforcing the soil mass should be taken to reduce the load acting on lining structure.

  6 Increased earth pressure induced by increased water content of soil caused by exfiltration in the operating period or other reasons shall be considered.

7.1.3 The calculation of soil tunnel lining shall meet the following requirements:

  1 For the calculation of shotcrete or composite lining consisting of bolt-shotcrete and reinforced concrete, the reinforced concrete lining may be designed as per a bearing structure, and calculated by structural mechanics methods. The bolt-shotcrete support may be estimated by the method given in GB 50086 or finite element method, and amended using experience derived from similar projects and construction monitoring results.

  2 For shotcrete or bolt shotcrete lining, the allowable values of relative convergence for perimeter and crown subsidence shall be determined by structural analysis based on the distribution of groundwater, soil conditions and construction monitoring results. In the absence of measured data, the values may be determined according to SL 377.

3 For the calculation of reinforced concrete lining, the internal water pressure may be assumed to be only borne by the reinforced concrete lining, and the joint action of reinforced concrete lining and soil may be ignored.

7.2 Joints, Seepage Control and Waterstops for the Linings of Soil Tunnels

7.2.1 For a soil tunnel, a circumferential deformation joint should be provided every 6 m to 12 m along the alignment, and the circumferential joints in the invert, side and crown shall not be staggered. Reliable seepage control and waterstop measures should be adopted for deformation joints.

7.2.2 For longitudinal construction joints in a lining, surface roughening shall be performed, and waterstops provided. The construction sequence shall start with the lining of the invert, and be followed by the lining of side walls and crown, and the reverse construction sequence shall be avoided.

7.2.3 For a soil tunnel, reliable enclosed waterstops shall be used for longitudinal construction joints and circumferential deformation joints.

7.2.4 For a collapsible loess tunnel, in addition to conforming to Article 7.2.3, flexible waterstops of integral seal type shall be provided between soil supports and lining or within the lining structure, and the impermeability of lining structure concrete should not be lower than W8. Where a collapsible loess tunnel is connected with a rock tunnel, the lining length of the soil tunnel with flexible waterstops extended into the rock tunnel shall conform to Article 7.1.1. In addition, waterproofing curtain shall be provided at the junction as well.

7.2.5 The joints in linings for soil tunnels shall conform to Section 9.7.

8 Design of Tunnels in Poor Ground

8.0.1 Poor ground refers to areas:

——Where a tunnel crosses major geological structures and special construction and support measures are required to ensure the stability of surrounding rock;

——With high pressure groundwater or strong water recharge areas where large inflow water may occur;

——With high in situ stress and/or where rock bursts may occur;

——Containing hazardous gases and radioactive elements;

——With developed karst caves or underground rivers;

——With soft rock strata, swelling rock strata, soil layer, sand layer, quicksand layer or landslide accumulation layer;

——Prone to erosion or seepage deformation/failure due to seepage;

——With high ground temperature.

8.0.2 The support design shall meet the following requirements:

1 According to geological prediction/forecast or advance exploration results, support or reinforcement design of surrounding rock shall be conducted before excavation with reference to the experience derived from similar projects and through necessary analysis. In case of unexpected circumstances, an emergency plan should be proposed as well.

2 According to exposed geological conditions, safety monitoring and test data during construction process, support parameters shall be verified and adjusted, or support design shall be modified timely.

3 The effect of initial support shall be analyzed timely. According to the stability of surrounding rock, the necessity of strengthening support or multiple rounds of support, and the appropriate timing for lining construction shall be studied.

4 Bolt-shotcrete support design shall comply with SL 377. Structural mechanics methods may be used in the structural calculation of other types of supports.

8.0.3 The lining design shall meet the following requirements:

1 According to geological conditions, effects of various treatment measures taken before lining construction and the state of surrounding rock deformation and/or displacement, the external loads possibly borne by the lining structure shall be determined with reference to the experience derived from similar projects and through necessary analysis.

2 Physical and mechanical indexes of surrounding rock adopted in design shall be determined through necessary tests and with reference to similar projects experience.

3 Tunnel shape and lining type favorable to the structural stress and stability of surrounding rock shall be selected through technical and economic comparison in light of geological and construction conditions.

4 Structural mechanics methods may be used for lining structural calculation when the elastic resistance of surrounding rock is ignored. The calculation method shall conform to Article 9.3.3 and be determined with reference to experience derived from similar projects when the elastic resistance of surrounding rock is considered.

8.0.4 Where collapse failure of surrounding rock may occur, the tunnels shall be constructed using the New Austrian Tunneling Method (NATM), and the following requirements shall be met:

1 Specific construction planning should be made.

2 Specific construction techniques should be proposed, including blasting parameters, footage, procedure, deformation monitoring, field tests, support technology, etc.

3 Diversion and drainage of groundwater should be well designed.

4 The stability of surrounding rock shall be timely judged on the basis of information feedback. Subsequent construction measures shall be determined based on the judgment.

8.0.5 For tunnel portions with relatively large inflow of water, taking into account of geological conditions, sources, quantities and environmental impacts of water inflow, engineering measures shall be proposed to prevent the inflow of water from causing failure of surrounding rock, in line with the principles of cutting off the water sources, diverting and draining water inflow, and reducing rock permeability.

8.0.6 In high in situ stress areas where rock bursts occur, the orientation and cross-sectional shape of the tunnel portion, excavation procedure, support modes, prerelease of rock pressure, etc. shall be studied in light of the magnitude and direction of in situ stress, structure and lithology of surrounding rock, and frequency, intensity and extent of rock bursts. The effects of supports shall be monitored closely. Lining construction shall be carried out after the state of deformation (displacement) of surrounding rock is mostly stable.

8.0.7 For tunnel portions through areas containing hazardous gases, measures including insulation, closure, ventilation and exhausting, etc. should be studied and selected in order to control and reduce the impacts of hazardous gases, according to the sources, distribution and connection of hazardous gases. Specific ventilation and air exchange facilities may be provided for tunnels with a great length or excessive concentration of hazardous gases. Bolt-shotcrete should not be used as permanent lining in hazardous gas areas.

8.0.8 For tunnels through karst areas, according to location, distribution, size and filling status of karst caves, stability and water quantities of surrounding rock (rock boundary), the following treatment measures shall be taken:

1 For seepage water and dripping water escaping from rock, flowing water (underground river) in karst caves and pore water in filling materials, comprehensive treatment measures such as drainage, cut, blocking, and prevention with priority given to drainage should be taken according to water quantity, type and source.

2 Treatment measures such as backfilling concrete, backfill grouting and consolidation grouting may be taken for relatively small scale karst caves or small size karst caves which are not connected with tunnels.

3 Measures such as providing partition, emplacing supporting structure for crossing, setting specialized foundation, and locally modifying alignment may be taken for large sized, water-rich and fillings-abundant karst caves, according to the location and distribution of caves.

8.0.9 For tunnel portions in strata with soft and swelling rock, the stress/strain relationship of soft rock, swelling ratio and pressure of swelling rock shall be studied first, based on geological exploration and test results. And appropriate support measures, section enclosing modes, enclosing time, and appropriate lining structure and lining construction time shall be determined with reference to the experience derived from similar projects and necessary calculation and analysis.

8.0.10 For tunnel portions in areas with large faults, unloading zones, fractured zones, jointed/fractured zones prone to erosion and seepage deformation/failure under seepage action, seepage control and waterstop of the lining structure shall be strengthened and specific design conducted when necessary.

8.0.11 For tunnels in poor ground, grouting, water control and drainage, waterstops for construction joints and structure joints, and safety monitoring shall be designed based on geological conditions and lining type.

9 Tunnel Supports and Linings

9.1 General

9.1.1 Tunnel supports shall be used to stabilize the surrounding rock or provide sufficient stability time for construction.

9.1.2 Tunnel linings shall have one or more of the following functions:

—To reinforce the surrounding rock, and bear loads jointly with the surrounding rock and supports;

—To smooth the surrounding rock surface;

—To improve the impermeability of the surrounding rock;

—To protect the rock against water erosion and scouring;

—To prevent adverse effects of temperature, humidity, atmosphere and other factors on the surrounding rock.

9.1.3 Tunnel support and lining design shall give full play to the self stabilization and bearing capacity of surrounding rock.

9.1.4 Support types include bolt, bolt-shotcrete, steel arch, mesh reinforced shotcrete, reinforced concrete, etc. Specific support type shall be determined by analysis or from projects experience in accordance with engineering geology, hydrogeology, sectional sizes and construction methods, etc.

9.1.5 The tunnel lining type shall be determined through technical and economic comparison with comprehensive consideration of the tunnel shape and size, internal water pressure, operating conditions, geological conditions, seepage control requirements, support effects, lining requirements, construction methods and other factors.

9.1.6 Lining types include bolt-shotcrete linings, unreinforced concrete linings, reinforced concrete linings and prestressed concrete linings (mechanical or grouting type), etc.

9.1.7 Linings shall meet requirements of both structure stability and seepage control. For tunnel linings with strict seepage control requirements or severe exfiltration risks due to poor impermeability of the surrounding rock, effective seepage control measures shall be adopted. Prestressed concrete linings or steel linings should be adopted if necessary.

9.1.8 The limit states method shall be used in the design of concrete lining structures for hydraulic tunnels. Given specified material strength and selected load values, the design shall be carried out using the expression of safety factors based on multi-factor analysis, and shall comply with SL 191.

9.2 Loads and Load Combinations

9.2.1 Classification of loads acting on lining is shown in Table 9.2.1.

Table 9.2.1 Load classification

Load classification	Loads
Permanent loads	Self-weight of lining, rock pressure, in situ stress, prestress, etc.
Variable loads	Internal water pressure, external water pressure, grouting pressure, construction loads and temperature load, etc, under normal operating conditions. The internal water pressure includes hydrostatic and hydrodynamic pressure (water hammer pressure, fluctuating pressure, time-average pressure of gradually varied flow)
Accidental loads	Internal and external water pressure in case of check flood level, and seismic load

9.2.2 Bearing capacity limit states design shall be performed in accordance with fundamental load combination and accidental load combination. The load combination shall meet the following stipulations:

　　1　Fundamental load combination shall be a combination of permanent and variable load effects.

　　2　Accidental load combination shall be a combination of permanent, variable load effects and one of the accidental load effects.

9.2.3　Checking of serviceability limit states shall be performed based on the load effects combination using characteristic values of permanent and variable loads.

9.2.4　The rock pressure acting on the lining shall be determined based on the rock conditions, shape and size of the cross section, construction methods, and support effects, and:

　　1　May be neglected for the surrounding rock with good self-stabilization conditions, whose state of deformation is quickly stable after excavation.

　　2　May be appropriately reduced when support measures are taken to make the surrounding rock basically stable or stable during excavation of tunnels.

　　3　Should be calculated as per the weight of the overlying rock mass, and adjusted based on support measures to be taken during construction for a shallow tunnel where stable ground arches cannot form.

　　4　May be determined according to the gravity of unstable blocks in the surrounding rock with blocky, medium thick or thick layer structure.

　　5　May be calculated by Equations (9.2.4-1) and (9.2.4-2) for a thin-layered or fractured rock:

　　For vertical direction

$$q_v = (0.2 \text{ to } 0.3) \gamma_R b \quad (9.2.4-1)$$

For horizontal direction

$$q_h = (0.05 \text{ to } 0.10) \gamma_R h \qquad (9.2.4-2)$$

Where:

q_v = vertical uniform rock pressure, kN/m^2;

q_h = horizontal uniform rock pressure, kN/m^2;

γ_R = unit weight of rock mass, kN/m^3;

b = excavated tunnel width, m;

h = excavated tunnel height, m.

6 May be reduced appropriately based on the surrounding rock conditions for tunnels excavated by TBM.

7 Shall be specifically studied for surrounding rocks with special properties such as rheology or swelling, which may generate deformation pressures on lining structures.

9.2.5 The tunnel internal water pressure shall be studied, with consideration of the tunnel layout, operating conditions, characteristic water levels at inlet and outlet, as well as the project-specific conditions and various operating conditions, and then determined as the maximum internal water pressure that may possibly occur (including hydrodynamic water pressure).

9.2.6 External water pressure acting on the lining may be determined in accordance with Annex C.

9.2.7 The adverse effects on lining such as stress caused by temperature variation, concrete shrinkage and expansion, and non-prestressed grouting shall be minimized through construction and structural measures. For high ground temperature zone, temperature stress shall be studied specially.

9.2.8 Seismic loads shall be determined in accordance with SL 203.

9.2.9 The construction loads may be determined according to the mechanical action during construction and maintenance.

9.3 Unreinforced- and Reinforced-Concrete Linings

9.3.1 The thickness of unreinforced- and reinforced-concrete linings shall be analyzed and determined in light of strength, impermeability and structure, in combination with construction methods, and the following requirements shall be met:

1 The thickness of concrete linings with one-layer reinforcement should not be less than 0.3 m; and that of concrete linings with two-layer reinforcement should not be less than 0.4 m.

2 The strength, impermeability and freezing-thawing resistance of unreinforced- and reinforced-concrete linings shall conform to SL 191. The index of abrasion- and erosion-resistance may be selected according to DL/T 5207.

3 Seepage control may not be required for unreinforced concrete lining with the sole purpose of smoothing the rock surface.

9.3.2 For unreinforced- and reinforced-concrete linings, the bearing capacity calculation for limit state shall be carried out, and whether or not the checking of serviceability limit states shall be performed depends on its function, requirements for impermeability and durability, as well as the impermeability of the surrounding rock. In case of checking the serviceability limit states, the crack width may be calculated according to Annex D, and the maximum allowable crack width shall comply with SL 191.

9.3.3 A proper method may be selected in the calculation of tunnel lining structures according to lining type, load features, surrounding rock conditions, construction methods and procedures, etc. And the following shall be complied with:

1 The finite element method should be adopted for high-pressure tunnels or important hydraulic tunnels.

2 Elastic mechanics analytical method may be adopted for circular pressure tunnels with relatively homogeneous surrounding rock or rock cover in accordance with Clauses 2 to 4 of Article 4.1.4, and the elastic resistance of the surrounding rock shall be considered in calculation. When the thickness of the circumferential surrounding rock of the tunnel is less than 3 times the excavation diameter, the resistance shall be determined through justification.

3 Boundary numerical method should be adopted for circular free-flow tunnels and tunnels with other cross-sectional shapes.

4 The finite element method may be adopted for the multiple parallel tunnels and the interaction between the tunnels shall be considered.

5 When the size of tunnel sections is large and internal and external water heads are high, after justification, the lining may be considered as permeable lining in calculation.

9.3.4 The structural calculation of fabricated concrete linings should be performed by one or more of the methods such as modified routine method, ring model with equivalent stiffness, lining boundary method, and finite element method. The types of joints and connection between precast segments shall be considered as well.

9.3.5 For linings bearing asymmetric loads, specific calculations may be required based on topographical and geological conditions.

9.4 Prestressed Concrete Linings

9.4.1 For pressure tunnels with high seepage control requirements or with overlying rock mass dissatisfying the requirements of hydraulic jacking, prestressed concrete linings may be adopted.

9.4.2 Prestressed linings may be divided into grouting type and mechanical type according to how the prestress is applied. The type of prestressed concrete linings should be selected based on geological conditions and operating requirements. Mechanically prestressed concrete linings may be used in all surrounding rocks, while the grouting type may be used just in hard rocks or rocks strong enough to bear the grouting pressure after treatment.

9.4.3 Circular sections shall be adopted for prestressed linings, and the lining structure shall meet the following requirements:

1 Under the combined actions of internal water pressure, prestress and other loads, the tensile stress of linings shall be less than the allowable tensile stress of concrete.

2 Without internal water pressure, under the combined actions of prestress and other loads, the compressive stress of linings shall be less than the allowable compressive stress of concrete.

3 The lining thickness shall be determined based on the calculation of different load combinations under various operating conditions. The minimum thickness should not be less than 0.6 m for mechanical type, nor 0.3 m for grouting type.

9.4.4 For prestressed concrete linings, the adopting of design indexes of material properties for concrete and reinforcing bars/tendons and the strength safety factors of prestressed lining structures, and the calculation of stress loss of steel cable shall conform to SL 191.

9.4.5 The calculation of bearing capacity limit state and the checking of serviceability limit state shall be carried out for prestressed concrete linings.

9.4.6 The relevant parameters and construction technology of prestressed concrete linings shall be determined through tests.

9.4.7 The smooth blasting method should be adopted for prestressed concrete lining tunnels. When large overbreaks occur on an excavation section, backfilling should be carried out first.

9.4.8 Mechanically prestressed linings are divided into two types:

bonded post-tensioned prestressed and unbonded post-tensioned prestressed linings. Design priority should be given to the unbonded post-tensioned prestressed linings.

9.4.9 Mechanically post-tensioned prestressed steel/tendons should be set at the outer edge of the lining. The spacing shall be determined by calculation, but should not be more than 0.5 m. Measures shall be taken to reduce the friction coefficient between tendons and hole walls. The locations of anchorage devices should be staggered.

9.4.10 For mechanically prestressed concrete linings, grouting shall comply with the following:

    1 Contact grouting shall be carried out covering all interface between lining and surrounding rock.

    2 For bonded post-tensioned prestressed linings, holes shall be grouted and tensioning slots backfilled in time after tendons have been tensioned.

9.4.11 For grouting prestressed concrete linings, grouting parameters shall be determined through field tests based on design requirements, and grouting procedures should be as follows:

    1 Consolidation grouting of surrounding rock.

    2 Injecting high-pressure water between linings and surrounding rock until they are completely separated.

    3 High-pressure grouting between linings and surrounding rock.

9.5 Unlined and Bolt-Shotcrete Lined Tunnels

9.5.1 For unlined or bolt-shotcrete lined tunnels, apart from the stability requirements of the surrounding rock, one of the following requirements shall be met:

    —No exfiltration. The surrounding rock is almost impervious after treatment, or external water pressure is higher than internal water pressure;

—No adverse consequences of exfiltration. Long-term exfiltration will neither pose any threat to the stability of the rock mass or hillside, or the safety of structures nearby, nor cause any environmental damage.

9.5.2 Bolt-shotcrete linings should not be adopted for tunnel portions:

—With large-area water inflow for a long time;

—With highly corrosive water;

—In swelling strata;

—With special requirements.

9.5.3 In the early design stage of bolt-shotcrete linings, the lining types and parameters may be selected preliminarily by referring to SL 377 and GB 50086, and with consideration of the engineering geological conditions, tunnel sizes and purposes, and service life of the project. In the construction design stage, according to the exposed geological conditions, the classification of the surrounding rock shall be modified and the types and parameters of bolt-shotcrete linings shall be adjusted.

9.5.4 The finite element method, elastoplastic numerical method or approximate analysis method should be adopted in checking the integral stability of the surrounding rock. The wedge limit equilibrium method may be adopted in checking the stability of the surrounding rock with the likelihood of localized failure.

9.5.5 The portal of an unlined or bolt-shotcrete lined tunnel shall be reinforced. The length of the reinforced portion should not be less than:

1 The length of unloading zone and intensively weathered zone behind the portal.

2 2 to 3 times the tunnel diameter or width.

9.5.6 The bottom of an unlined or bolt-shotcrete lined tunnel shall be leveled by cast-in-place concrete with a thickness not less than 0.2 m.

9.5.7 Rock traps shall be provided in the unlined or bolt-shotcrete lined water conveyance tunnels of hydropower stations. The locations, volume and number of rock traps may be determined with comprehensive consideration of surrounding rock conditions, tunnel length, hydraulic conditions, debris cleaning frequency and methods, and the following requirements shall be met:

   1 The flow disturbance in the tunnel cross section and rock traps should be reduced.

   2 Measures to prevent sand and stone from moving in longitudinal direction should be adopted in the traps.

   3 Model tests should be carried out for rock traps of important projects.

   4 Accessibility for maintenance shall be considered.

9.5.8 The progress of bolt-shotcreting should closely follow that of excavation. The average unevenness of coat surface should not exceed 0.15 m.

9.5.9 The allowable flow velocity should not exceed 8 m/s for a bolt-shotcrete lined tunnel nor 12 m/s for a tunnel with temporary water flow.

9.5.10 The shotcrete strength grade shall not be lower than C20. The bonding strength between the shotcrete coat and surrounding rock should not be lower than 1.0 MPa for surrounding rock masses of Classes Ⅰ and Ⅱ, nor 0.8 MPa for Class Ⅲ.

9.5.11 For the surrounding rock with the likelihood of large plastic deformation due to excavation or located in rock burst-prone and high in situ stress areas, steel fiber shotcrete or synthetic fiber shotcrete should be adopted. The surface of steel fibre shotcrete shall be coated with either an ordinary shotcrete layer of no less than 30 mm in thickness or a cement mortar layer of no less than 10 mm in thickness, and the strength grade of the layer shall not be lower than that of the steel fiber shotcrete. The steel fiber shotcrete shall conform to SL 377 and GB 50086, and the synthetic fiber shotcrete shall conform to GB/T 21120.

9.5.12 The bearing capacity of the rock bolts/tendons that reinforce the surrounding rock shall be calculated in accordance with the stipulations of SL 377. Rock bolts/tendons shall be installed in the direction favorable to bearing loads, and shall extend into the stable surrounding rock for a sufficient anchorage length.

9.5.13 For the surrounding rock with poor integral stability, pattern bolts should be adopted and arranged as follows:

1 The rock bolts at cross section should be perpendicular to the plane of major discontinuities, and may be perpendicular to the peripheral line of the tunnel where the plane of major discontinuities is not obvious.

2 The rock bolts should be arranged on the surrounding rock surface in the shape of diamond, rectangle and square.

3 The spacing of rock bolts should not be more than one-half their length. The spacing of rock bolts in surrounding rocks of Classes Ⅳ and Ⅴ should range from 0.5 m to 1.0 m, and shall not exceed 1.5 m.

9.5.14 Bolt-shotcrete linings with mesh shall meet the following stipulations:

1 For steel mesh, longitudinal and circumferential reinforcing bars should be 6 mm to 12 mm in diameter, and spaced 0.15 m to 0.3 m apart.

2 The steel mesh and rock bolts should be joined by welding.

3 Every other intersection point of steel mesh shall be well welded or bound.

4 The thickness of shotcrete cover should not be less than 50 mm.

9.6 Design of Reinforced Concrete Lined Bifurcation

9.6.1 Reinforced concrete lined bifurcation should be located in the rock masses of Class I or Class II, and shall conform to Article 4.1.4. It may be located in the Class III rock mass after justification, but shall not be located in the rock masses of Classes IV and V.

9.6.2 A reinforced concrete lined bifurcation and tunnel portions immediately upstream and downstream of this bifurcation shall satisfy the requirements of minimum rock cover, hydraulic jacking resistance and seepage control. The field tests of in situ stress and/or physical and mechanical parameters for surrounding rock shall be carried out if necessary.

9.6.3 Such factors as project layout, operating requirements, hydraulic conditions, construction methods shall be comprehensively considered in determining the type, shape and size of a reinforced concrete lined bifurcated tunnel. The angle between two branches should be in the range of 45° to 60°. The bifurcation shall be provided with a smooth transition. The connection contour of the bifurcation should be rounded rather than broken.

9.6.4 The structural design of a reinforced concrete lined bifurcated tunnel shall meet the following:

1 For a reinforced concrete lined bifurcated tunnel located in impervious or slightly permeable rock masses of Class I or Class II, the internal water pressure may be ignored, and reinforcement may be determined with reference to the experience derived from similar projects and detailing requirements.

2 The structural calculation of the bifurcated tunnel with reinforced concrete linings may conform to Article 9.3.3.

3 For important projects, the finite element method shall be employed in calculation.

9.6.5 Backfill grouting and consolidation grouting shall be carried out for the bifurcated tunnel, conforming to Section 10.1.

9.7 Joints in Linings

9.7.1 For unreinforced- and reinforced-concrete linings, permanent joints shall be provided at the intersection of shaft and tunnel, the inlet and outlet, the positions where geological conditions changes significantly or large relative displacement may occur, and corresponding seepage control measures shall be taken.

9.7.2 For tunnel portions with relatively geologically homogeneous surrounding rock, it is possible to set construction joints only. The spacing of construction joints may be determined by such factors as construction method, concrete placing capability, and atmospheric temperature variation. The spacing should be 6 m to 12 m. The circumferential joints in the invert, sides and crown shall not be staggered.

9.7.3 For the circumferential construction joints without seepage control requirement, it is not necessary for distributed reinforcement to pass through the joint. Waterstops are not required. For those with seepage control requirement, necessary joint treatment measures shall be taken according to specific conditions.

9.7.4 Longitudinal construction joints shall be set at the locations with low tensile stress and shear stress. For joints without seepage control requirement, waterstops are not required. For joints with seepage control requirement, necessary joint treatment measures shall be taken according to specific conditions. Where lining construction starts from the side walls and crown, the joints in abutments shall be properly treated.

9.7.5 No joints shall be set at the connection of reinforced concrete lining and steel lining, and the overlap length shall not be less than 1.0 m.

9.8 Design of Water-Retaining Plugs

9.8.1 For plugs of hydraulic tunnels which are connected to the reservoir directly, their classes and requirements for stability and seepage control shall be consistent with those for water-retaining structures. The classes of plugs of adits shall be the same as those of the hydraulic tunnels to which they are connected.

9.8.2 The locations of plugs shall be determined by analysis of engineering geological and hydrogeological conditions of surrounding rocks, existing supports or linings, arrangement of adjacent structures, and operating requirements.

9.8.3 The shapes and lengths of plugs shall be determined by comprehensive analysis of water pressure acting on plugs, geological conditions, construction methods, materials of plugs, operating requirements, as well as construction duration.

9.8.4 A plug shall be set on the grout curtain line of a water-retaining structure when the axis of a river diversion tunnel passes through such curtain.

9.8.5 Plugs shall be made of concrete, whose indexes, such as strength and permeability shall be determined according to SL 191. Slightly expansive concrete may be adopted, and the type and dosage of the expansive admixture should be determined through tests.

9.8.6 For mass concrete plugs, thermal control measures should be considered.

9.8.7 Plugs shall be so designed as to be under bearing capacity limit state, and the calculation principles, loads, load effect combinations and the related parameters shall conform to SL 191.

9.8.8 The sliding safety factor for plugs calculated as per shear breaking strength shall not be less than 3.0.

9.8.9 The sliding stability of plugs shall be calculated by Equations (9.8.9-1) through (9.8.9-3):

$$KS \leqslant R \qquad (9.8.9-1)$$
$$S = \sum P \qquad (9.8.9-2)$$
$$R = f' \sum W + C' \sum \lambda_i A_i \qquad (9.8.9-3)$$

Where:

K = Sliding safety factor calculated as per shear breaking strength;

S = Design values of load effects;

R = Design values of the bearing capacity of plugs;

$\sum P$ = Sum of tangential components of all loads borne by plugs on the sliding plane, kN;

$\sum W$ = Sum of normal components of all loads borne by plugs on the sliding plane, taking as positive for the downward, kN;

f' = Friction coefficient of shear breaking strength between concrete and surrounding rock or between concretes;

C' = Cohesion of shear breaking strength between concrete and surrounding rock or between concretes, kPa;

A_i = Contact area of bottom and sides of a plug with surrounding rock/concrete, except that of crown, m^2;

λ_i = Effective area coefficient of the contact surface of bottom and sides of a plug with surrounding rock/concrete except that of crown, λ is taken as 1 for bottom, and determined according to project-specific conditions for sides.

9.8.10 If a grouting gallery is provided in the plug, the length of the solid part in front of the galley shall be checked.

9.8.11 For plugs bearing high internal water pressure, the finite element analysis shall be carried out.

9.8.12 The grouting of plugs shall conform to Article 10.1.7.

9.8.13 When a plug overlaps with the tunnel lining, the overlap length shall not be less than 2 m. Circumferential waterstops shall be installed within the overlap.

10 Grouting, Seepage Control and Drainage of Tunnels

10.1 Grouting

10.1.1 Backfill grouting must be carried out between the surrounding rock and the concrete lining (unreinforced or reinforced) crown or the plug top.

10.1.2 The extent, hole spacing, row distance, grouting pressure and grout concentration to be applied in backfill grouting shall be determined by the analysis of lining type, operating conditions and construction methods. Backfill grouting should be performed either in the whole crown or only within the central angle range of 90° to 120°. The grouting of other portions is dependent upon the placing of lining. Both holes and rows should be spaced 3 m to 6 m. The grouting pressure may be 0.2 MPa to 0.3 MPa for unreinforced concrete linings, and 0.3 MPa to 0.5 MPa for reinforced concrete linings. Grouting hole shall extend into the surrounding rock for at least 0.1 m.

10.1.3 Low pressure should be adopted in backfill grouting of soil tunnel. When flexible waterstops are provided between supports and linings, grout pipes shall be embedded prior to lining placement, and shall neither damage flexible waterstops nor pass through supports.

10.1.4 Cement blocks formed by backfill grouting shall be able to transmit resistance force.

10.1.5 Consolidation grouting of surrounding rocks shall be determined through technical and economical comparison based on geological and hydrogeological conditions of the tunnel, types of linings, influences of construction on surrounding rocks, and operating requirements. The rows of consolidation grouting holes should be spaced 2 m to 4 m. At least 6 holes should be arranged symmetrically for each row. The grouting depth into the surrounding rock shall be determined by the analysis of the surrounding rock conditions, and should not be less than half of the tunnel diameter or width. The grouting pressure may be 1 to 2 times the internal water pressure.

10.1.6 For consolidation grouting with special requirements or high pressure, parameters may be determined by similar projects experience or in situ tests.

10.1.7 The grouting of tunnel plugs shall meet the following:

1 Consolidation grouting of surrounding rocks should be carried out at the tunnel plug. The grouting parameters may be determined after studying rock conditions, water head acting on the plug, and types of plugs. Consolidation grouting at diversion tunnel plug should be executed within the plug galley.

2 Backfill, joint and contact grouting shall be carried out around the plug.

3 The grouting parameters such as grouting pattern, grouting pressure, and grout concentration shall be determined by analysis of geological and hydrogeological conditions and types, operating condition and construction methods of the plug.

10.1.8 The grouting materials shall be selected according to geological and hydrogeological conditions and operating conditions of the tunnel. In the presence of corrosive groundwater, corrosion-resistant cement shall be adopted.

10.1.9 The grouting of hydraulic tunnels shall also conform to SL 62.

10.2 Seepage Control and Drainage

10.2.1 Seepage control and drainage measures shall be determined by comprehensive analysis of factors such as geological and hydrogeological conditions of surrounding rocks along the tunnel, the lining types, and requirements related to environmental protection, soil and water conservation and operation.

10.2.2 For free-flow tunnels, drain holes should be above the water surface. The spacing, row distance and depth of drain holes shall be determined by analysis of geological conditions and external water conditions. For developed fissures with rock debris in the surrounding rock, soft pervious pipes shall be provided in drain holes to avoid the rock debris being brought out with water.

10.2.3 When the external water pressure is the control factor in designing the lining of the pressure tunnel, appropriate drainage measures should be considered to reduce external water pressure.

10.2.4 Effective seepage control measures shall be adopted for the following parts of a hydraulic tunnel to avoid failure due to seepage of surrounding rocks and slopes.

——Portal of pressure tunnel;

——Rock mass between adjacent high pressure tunnel portions;

——Tunnel portions in poor ground or in Classes Ⅳ and Ⅴ rocks;

——Tunnel portions where the rock cover fails to meet the requirements of Article 4.1.4.

10.2.5 According to the topographical and geological conditions, drain holes and intercepting ditches shall be provided on the tunnel portal slope to form a reliable drainage system. Measures shall be taken on portal slopes to prevent erosion caused by surface runoff.

10.2.6 At the transition between reinforced concrete lining and steel lining of a high pressure tunnel, a circumferential grout curtain shall be provided at the end of the reinforced concrete lining or at the upstream end of the steel lining; and seepage rings shall be provided at the upstream end of the steel lining.

11 Tunnel Operation and Maintenance

11.0.1 Operation rules of a hydraulic tunnel shall be prepared according to its operating requirements, together with natural conditions, structural design, and relevant studies and tests.

11.0.2 The rules shall cover tunnel operation, emptying, inspection, maintenance, etc.

11.0.3 Necessary facilities and signs shall be provided according to management and maintenance requirements of the hydraulic tunnel.

Annex A Head Loss Calculation for Hydraulic Tunnels

A.1 Friction Losses

A.1.1 Friction losses shall be calculated by Equations (A.1.1-1) and (A.1.1-2):

$$h_f = \frac{Lv^2}{C^2 R} \qquad (A.1.1-1)$$

$$C = \frac{1}{n} R^{\frac{1}{6}} \qquad (A.1.1-2)$$

Where:

R = Hydraulic radius, m;

n = Roughness, see Table A.1.1.

Table A.1.1 Roughness n of pressure tunnels

No.	Tunnel surface	Roughness n		
		Mean	Max	Min
1	Unlined rock surface	—	—	—
	(1) Smooth blasting	0.030	0.033	0.025
	(2) Ordinary drilling and blasting	0.038	0.045	0.030
	(3) Full-face TBM excavation	0.017	—	—
2	Cast-in-place concrete lining with steel forms	—	—	—
	(1) Ordinary technique	0.014	0.016	0.012
	(2) Good technique	0.013	0.014	0.012

Table A.1.1 (continued)

No.	Tunnel surface	Roughness n		
		Mean	Max	Min
3	Shotcreted rock surface	—	—	—
	(1) Smooth blasting	0.022	0.025	0.020
	(2) Ordinary drilling and blasting	0.028	0.030	0.025
	(3) Full-face TBM excavation	0.019	—	—
4	Steel pipe	0.012	0.013	0.011

A.1.2 For unlined tunnels or shotcrete supported tunnels where the concrete lining is only used on floor or other positions, the roughness (taken as composite roughness n_0) may be calculated by Equation (A.1.2):

$$n_0 = n_1 \left[\frac{S_1 + S_2 \left(\frac{n_2}{n_1}\right)^{\frac{3}{2}}}{S_1 + S_2} \right]^{\frac{2}{3}} \quad (A.1.2)$$

Where:

n_0 = Composite roughness;
n_1 = Roughness of unlined surface;
n_2 = Roughness of concrete lining;
S_1 = Wetted perimeter of unlined surface;
S_2 = Wetted perimeter of concrete lining.

A.2 Form Losses

A.2.1 Form losses shall be calculated by Equation (A.2.1):

$$h_m = \xi \frac{v^2}{2g} \quad (A.2.1)$$

Where:

ξ = Form loss coefficient, see Tables A.2.1-1 to A.2.1-3.

Table A.2.1-1 Form loss coefficient ξ

No.	Position	Shape	Form loss coefficient ξ	Notes
1	Intake		0.5	v—velocity at uniform section of a pipe
			0.25	
			0.2 ($r/d < 0.15$) 0.1 ($r/d \geq 0.15$)	
2	Trash rack		$\beta \left(\dfrac{s}{b}\right)^{\frac{4}{3}} \sin\alpha$	β—bar shape coefficient, see Table A.2.1-2; s—bar width; b—bar spacing; α—inclination angle of rack; v—mean velocity in front of trash rack
3	Gate slot		0.05 – 0.20 (may be taken as 0.10)	v—mean velocity at upstream and downstream of the slot

Table A.2.1-1 (continued)

No.	Position	Shape	Form loss coefficient ξ	Notes
4	Rectangular to circular (smoothly contracted)		0.05	v—mean velocity at transition section, $v = \dfrac{v_1 + v_2}{2}$
5	Circular to rectangular (smoothly contracted)		0.1	
6	Conical expansion		ξ; see figure A.2.1-1	v_1 shall be taken
7	Conical contraction		ξ_d see figure A.2.1-2	v_1 shall be taken

Table A.2.1-1 (continued)

No.	Position	Shape	Form loss coefficient ξ	Notes
8	Circular bend		$\left[0.131+0.1632\times\left(\dfrac{D}{R}\right)^{\frac{7}{2}}\right]$ $\times\left(\dfrac{\theta}{90°}\right)^{\frac{1}{2}}$	D—tunnel diameter; R—bend radius; θ—deflection angle
9	Outlet		$\left(1-\dfrac{A_1}{A_2}\right)^2$ (shall be taken as 1 where the water in downstream channel is deep)	A_1—sectional area in front of the outlet; A_2—sectional area behind the outlet; v—velocity in front of outlet

Table A.2.1-1 (continued)

No.	Position	Shape	Form loss coefficient ξ	Notes
10	Right angle bifurcation		0.1	—
			1.5	—
11	Symmetrical Y-shaped bifurcation		0.75	Without a conical pipe section
			0.5	With a conical pipe section

Table A. 2. 1 - 1 (continued)

No.	Position	Shape	Form loss coefficient ξ	Notes
12	T-shaped bifurcation		See Equations (A. 2. 2 - 1) through (A. 2. 2 - 4) for dividing flows	—
			See Equations (A. 2. 2 - 5) through (A. 2. 2 - 8) for combining flows	—
13	Butterfly valve		See Table A. 2. 1 - 3	D—disc diameter; t—disc thickness

Table A. 2. 1 - 2 Bar shape coeffcient β

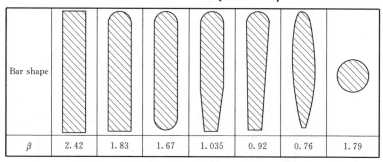

Bar shape							
β	2.42	1.83	1.67	1.035	0.92	0.76	1.79

Table A. 2. 1 - 3 Relationship between ξ and t/D ratio under full opening of butterfly valve

t/D	0.1	0.15	0.2	0.25
ξ	0.05 - 0.10	0.10 - 0.16	0.17 - 0.24	0.25 - 0.35

Note: ξ shall be taken as 0.2 under full opening of butterfly valve in the absence of relevant data.

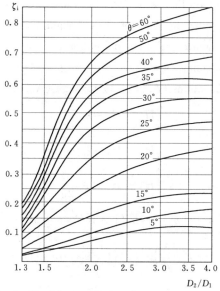

Figure A. 2. 1 - 1 Head loss coefficient ξ_i in conical expansion ($\theta < 60°$)

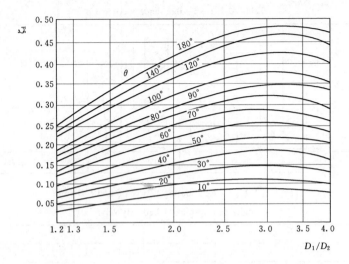

Figure A.2.1-2　Head loss coefficient ξ_d in conical contraction

A.2.2 Dividing flow and combining flow in T-shaped bifurcation tunnel are shown on Figure A.2.2, form losses may be preliminarily estimated according to the following requirements and shall be determined through model tests and numerical analysis if necessary.

1 For dividing flows ($Q_1 = Q_2 + Q_3$), form losses may be calculated by Equations (A.2.2-1) through (A.2.2-4):

$$\begin{cases} H_2 - H_1 = \xi_2 \dfrac{v_1^2}{2g} \\ H_3 - H_1 = \xi_3 \dfrac{v_1^2}{2g} \\ H_3 - H_2 = \xi_{32} \dfrac{v_1^2}{2g} \end{cases} \quad \text{(A.2.2-1)}$$

$$\xi_2 = -0.95(1-q_2)^2 - q_2^2 \left(1.3\cot\dfrac{\theta}{2} - 0.3 + \dfrac{0.4 - 0.1\psi}{\psi^2}\right)$$
$$\times \left(1 - 0.9\sqrt{\dfrac{\rho}{\psi}}\right) - 0.4\left(1 + \dfrac{1}{\psi}\right)\cot\dfrac{\theta}{2}(1-q_2)q_2$$
$$\text{(A.2.2-2)}$$

$$\xi_3 = -0.58q_2^2 + 0.26q_2 - 0.03 \quad \text{(A.2.2-3)}$$

$$\xi_{32} = (1-q_2)\left\{0.92 + q_2\left[0.4\left(1+\frac{1}{\psi}\right)\cot\frac{\theta}{2} - 0.72\right]\right\}$$
$$+ q_2^2\left[\left(1.3\cot\frac{\theta}{2} - 0.3 + \frac{0.4-0.1\psi}{\psi^2}\right)\right.$$
$$\left.\times\left(1 - 0.9\sqrt{\frac{\rho}{\psi}}\right) - 0.35\right] \qquad (A.2.2-4)$$
$$\rho = r/D$$
$$q_2 = Q_2/Q_1$$

Where:

H_1, H_2, H_3 = Total head at sections 1-1, 2-2 and 3-3;

v_1 = Mean velocity at section 1-1;

θ = Intersecting angle of main pipe and branch pipe;

ψ = The ratio of the cross-sectional area of the branch pipe to the cross-sectional area of the main pipe;

D = Diameter of main pipe;

r = Rounded radius at the intersection between branch pipe and main pipe;

Q_2 = Branch discharge;

Q_1 = Main pipe discharge before dividing. q_2 shall be positive when dividing.

2 For combining flows ($Q_1 + Q_2 = Q_3$), form losses may be calculated by Equations (A.2.2-5) through (A.2.2-8):

$$\begin{cases} H_2 - H_1 = \xi_2' \dfrac{v_3^2}{2g} \\ H_3 - H_1 = \xi_3' \dfrac{v_3^2}{2g} \\ H_3 - H_2 = \xi_{32}' \dfrac{v_3^2}{2g} \end{cases} \qquad (A.2.2-5)$$

$$\xi_2' = -0.95(1+q_2)^2 + q_2^2\left[1 + 0.42\left(\frac{\cos\theta}{\psi} - 1\right) - 0.8\left(1 - \frac{1}{\psi^2}\right)\right]$$

$$+ (1-\psi)\left(\frac{\cos\theta}{\psi} - 0.38\right)\right] \qquad \text{(A.2.2-6)}$$

$$\xi'_3 = q_2^2 \left[2.59 + (1.62 - \sqrt{\rho})\left(\frac{\cos\theta}{\psi} - 1\right) - 0.62\psi\right]$$
$$+ q_2(1.94 - \psi) - 0.03 \qquad \text{(A.2.2-7)}$$

$$\xi'_{32} = (1 + q_2)\left[0.92 + q_2(2.92 - \psi)\right] + q_2^2\left[\left(1.2 - \sqrt{\rho}\right)\right.$$
$$\left. \times \left(\frac{\cos\theta}{\psi} - 1\right) + 0.8\left(1 - \frac{1}{\psi^2}\right) - (1-\psi)\frac{\cos\theta}{\psi}\right] \text{(A.2.2-8)}$$

$$q_2 = Q_2/Q_3$$

Where:

Q_2 = Branch discharge;

Q_3 = Main pipe discharge after combining. q_2 shall be negative when combining;

The definitions of other symbols are the same as above.

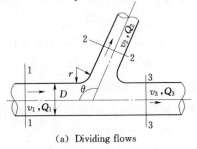

(a) Dividing flows

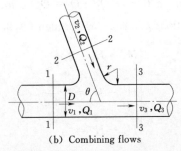

(b) Combining flows

Figure A.2.2 Sketch for form loss calculation of dividing flows and combining flows in T-shaped bifurcation

Annex B Cavitation Damage Prevention Design of High-Velocity Tunnels

B. 0. 1 When designing high-velocity hydraulic tunnels, the flow cavitation index σ shall be greater than the incipient cavitation index σ_i. The value of σ_i at important parts of high-velocity hydraulic tunnels shall be determined through tests in the construction design stage. For various hydraulic tunnels which are not often used (except for gate slot of diversion tunnels) and tunnel portions convenient for maintenance, the value of σ may be at least $0.85\sigma_i$. For high-velocity tunnels, the flow cavitation index σ along the tunnel may be calculated by Equations (B. 0. 1 – 1) through (B. 0. 1 – 3):

$$\sigma = \frac{P_0 + P_a - P_v}{\frac{1}{2}\rho_w v_0^2} \qquad (B.\,0.\,1-1)$$

$$P_a = \gamma_w (10.33 - \nabla/900) \qquad (B.\,0.\,1-2)$$

$$\rho_w = \frac{\gamma_w}{g} \qquad (B.\,0.\,1-3)$$

Where:

P_0 = Time-average dynamic water pressure at the calculated section, kPa; where the flow velocity exceeds 30 m/s, fluctuating pressure shall be considered;

P_a = Atmospheric pressure at the calculated section, kPa, it shall be estimated by Equation (B. 0. 1 – 2) for different elevation;

γ_w = Unit weight of water, kN/m³;

∇ = Height above sea level, m;

P_v = Vapor pressure of water, kPa, refer to Table B. 0. 1;

ρ_w = Density of water, kg/m³;

g = Acceleration of gravity, m/s²;

v_0 = Flow velocity at the calculated cross section, m/s, it may be the mean section velocity calculated according to the measured velocity distribution at the cross section.

Table B. 0. 1 Relationship between vapor pressure and temperature of water

Water temperature/ ℃	0	5	10	15	20	25	30	40
P_v/kPa	0.6	0.9	1.3	1.7	2.4	3.2	4.3	7.5

B. 0. 2 The requirements on control and processing of surface irregularities of the flow boundary shall be determined by the values of flow cavitation indexes as shown in Table B. 0. 2.

B. 0. 3 For energy dissipaters and erosion control structures in hydraulic tunnels or at outlets, where flow cavitation index is less than 0.3, aeration and cavitation prevention facilities shall be provided according to the following principles:

  1 Reasonable aeration type shall be selected and verified through large scale model tests.

  2 Enough quantity of air shall be provided to achieve the required concentration. The air concentration near the tunnel surface shall be greater than 4%.

  3 The protective aeration length shall be determined by discharge curve type and aeration structure type. The length may be 70 m to 100 m for a curved section and 100 m to 150 m for a straight one. The provision of multiple aeration and cavitation prevention facilities shall be considered for a long discharge tunnel.

B. 0. 4 Cavitation monitoring design shall be carried out for the high-velocity areas of Classes I and II discharge structures.

Table B.0.2 Surface irregularity control and processing standards

	>1.70	1.70–0.61	0.60–0.36	0.35–0.31	0.30–0.21		0.20–0.16		0.15–0.10		<0.10
Flow cavitation index σ					not required	required	not required	required	not required	required	
Aeration facilities	—	—	—	—	not required	required	not required	required	not required	required	Modify the design
Protrusion height/mm	≤30	≤25	≤12	≤8	<6	<25	<3	<10	Modify the design	<6	—
Smoothed slope — Positive slope	untreated	1/5	1/10	1/15	1/30	1/5	1/50	1/8	—	1/10	—
Smoothed slope — Side slope	untreated	1/4	1/5	1/10	1/20	1/4	1/30	1/5	—	1/8	—

Annex C Calculation Method and Reduction Coefficient of External Water Pressure

C. 0. 1 The external water pressure acting on unreinforced-, reinforced- or prestressed-concrete lining may be estimated according to Equation (C. 0. 1):

$$P_e = \beta_e \gamma_w H_e \qquad (C.0.1)$$

Where:

P_e = External water pressure acting on outer surface of lining structure, kN/m^2;

β_e = Reduction coefficient of external water pressure;

γ_w = Unit weight of water (taken as 9.81 kN/m^3), kN/m^3;

H_e = Water head between the groundwater level line and the center of the tunnel (taken as internal water pressure where exfiltration occurs), m.

C. 0. 2 For a concrete lined tunnel, the reduction coefficient of external water pressure may be selected from Table C. 0. 2 according to the groundwater conditions in surrounding rock in combination with the drainage measures to be adopted. For hydraulic tunnels equipped with drainage facilities, the external water pressure acting on the lining may be reduced according to the drainage effects and the reliability of the drainage facilities, and the reduction value may be determined empirically or by seepage calculation. Special studies shall be carried out for tunnels with complex geological and hydrogeological conditions, and high external water pressure.

Table C.0.2 Reduction coefficient of external water pressure β_e

Category	Groundwater descriptions	Groundwater impacts on surrounding rocks	β_e
1	Dry or moist tunnel wall	None	0 – 0.20
2	Seepage or drip along discontinuities	Weathering fillings in discontinuities, reducing shear strength of discontinuities, and softening soft rock	0.10 – 0.40
3	Large amount of dripping water, linear water flow or water jet along fissures or weak discontinuities	Causing argillization of fillings in discontinuities, reducing shear strength of discontinuities, and softening medium-hard rock masses	0.25 – 0.60
4	Dripping severely, and with small amount of inflow along weakness planes	Scouring fillings in discontinuities, speeding up the weathering of rock mass, softening and argillizing weakness zones such as faults, making them expanded and disintegrated, causing piping, and producing seepage pressure which can cause the failure of thin weakness zones	0.40 – 0.80
5	Severe inrush water, and with large amount of inflow along weakness zones such as faults	Scouring away fillings in discontinuities, disintegrating rock mass, and producing seepage pressure which can cause the failure of thick weakness zones and collapse of surrounding rocks	0.65 – 1.00

Notes: When combined with internal water, β_e is taken as a smaller value. Without internal water, β_e is taken as a larger value.

Annex D Calculation of Concrete Lining Crack Width

D. 0. 1 The checking of the serviceability limit state for reinforced concrete lining shall comply with SL 191. The maximum crack width may be calculated according to Equations (D. 0. 2 - 1) through (D. 0. 2 - 6), and the lining cracking limit shall be applied in accordance with SL 191.

D. 0. 2 For axially- or eccentrically compressed tunnel linings with $e_0 \leqslant 0.5H$, the checking of crack width calculation is not required. For axially tensioned, eccentrically tensioned or eccentrically compressed tunnel linings with $e_0 > 0.5H$, the maximum crack width with consideration of non-uniform distribution of crack width and impacts of long-term load actions may be calculated by Equations (D. 0. 2 - 1) through (D. 0. 2 - 6):

$$\omega_{max} = 2\left(\frac{\sigma_s}{E_s}\Psi - 0.7 \times 10^{-4}\right)l_f \qquad (D.0.2-1)$$

$$\Psi = 1 - \alpha_2 \frac{f_{tk}}{\mu \sigma_s} \qquad (D.0.2-2)$$

$$l_f = \left(60 + \alpha_1 \frac{d}{\mu}\right)\nu \qquad (D.0.2-3)$$

$$d = \frac{4A_s}{S} \qquad (D.0.2-4)$$

$$\mu = \frac{A_s}{1000H} \qquad (D.0.2-5)$$

$$\mu = \frac{A_s}{1000h_0} \qquad (D.0.2-6)$$

Where:

ω_{max} = The maximum crack width, mm, for Grade Ⅲ steel bars used as tensile reinforcement, the calculated crack width shall be multiplied by a factor of 1.1;

l_f = Average spacing between cracks, mm;

Ψ = Strain non-uniformity coefficient of longitudinal tensile reinforcement between cracks, when $\Psi < 0.3$, $\Psi = 0.3$;

α_1, α_2 = Coefficients, for axial tension and small-eccentricity tension, $\alpha_1 = 0.16$, $\alpha_2 = 0.60$; for large-eccentricity tension, $\alpha_1 = 0.075$, $\alpha_2 = 0.32$; for large-eccentricity compression, $\alpha_1 = 0.055$, $\alpha_2 = 0.235$;

σ_s = Stress in tensile reinforcement of lining under serviceability state, it may be calculated in accordance with Article D.0.3 with all loads taken as characteristic values in calculation;

d = Diameter of tensile reinforcement, mm; for reinforcement with different diameters, d is calculated by Equation (D.0.2-4); for small-eccentricity tension, d is the diameter of reinforcement on the side with greater reinforcement stress;

A_s, S = Total area (mm^2) and total circumference (mm) of tensile reinforcement within one linear meter in the tunnel;

μ = Ratio of tensile reinforcement, it is calculated by Equation (D.0.2-5) for axial tension and small-eccentricity tension (in which $A_s = f_i + f_o$), and by Equation (D.0.2-6) for large-eccentricity tension and large-eccentricity compression (in which $A_s = f_i$ or f_o);

h_0 = Effective thickness of lining, mm;

f_i = Inner layer reinforcement area in lining, mm^2;

f_o = Outer layer reinforcement area in lining, mm^2;

H = Thickness of lining, mm;

f_{tk} = Characteristic value of axial tensile strength of concrete, N/mm^2;

$\nu=$ Coefficient related to surface shape of tensile reinforcement, for deformed reinforcing bars, $\nu=0.7$, for plain reinforcement, $\nu=1.0$, and for cold drawn low carbon reinforcement wires, $\nu=1.25$;

$e_o=$ Eccentricity of axial force to centroid of section, mm.

D. 0. 3 The stress in tensile reinforcement σ_s may be calculated using elastic mechanics method or finite element method when checking crack width in normal cross-section of circular pressure tunnel. It may be calculated by Equation (D. 0. 3 - 1) for axial tension or small-eccentricity tension, by Equation (D. 0. 3 - 2) for large-eccentricity tension, or by Equation (D. 0. 3 - 3) for large-eccentricity compression, when checking crack width on normal cross-section of a circular free-flow tunnel or noncircular tunnel:

$$\sigma_s = \frac{N_1}{A_s} \qquad (D.0.3-1)$$

$$\sigma_s = \frac{N_1}{A_s}\left(\frac{e}{z}+1\right) \qquad (D.0.3-2)$$

$$\sigma_s = \frac{N_1}{A_s}\left(\frac{e}{z}-1\right) \qquad (D.0.3-3)$$

$$z = (0.93 - 5\mu)h_0 \qquad (D.0.3-4)$$

$$z = \left(0.8 + 0.1\frac{e_0}{H} - 5\mu\right)h_0 \qquad (D.0.3-5)$$

Where:

$N_1=$ Axial force calculated based on characteristic values of long-term load combinations (rock pressure and external water pressure beneficial to structures should be ignored), N;

$e=$ Distance between action points of axial force N_1 and resultant force of tensile reinforcement;

$z=$ Distance between action points of resultant force of tensile reinforcement and resultant force of compression zone, for large-eccentricity tension, z is calculated by Equation

(D.0.3－4), for large-eccentricity compression, z is calculated according to Equation (D.0.3－5) and not more than $(0.93-5\mu) h_0$.

Explanation of Wording

Words in this specification	Equivalent expressions in specific situations	Strictness of requirement
Shall	It is necessary	Required
Shall not	It is not allowed/permitted It is unacceptable	
Should	It is recommended It is advisable	Recommended
Should not	It is not recommended It is not advisable	
May	It is suitable It is desirable It is preferable	Permitted
Need not	It is unnecessary It is not required	